```
VG 93 .A7 K56 1990

Kinzey, Bert.

Colors & markings
```

COLORS & MARKINGS OF U.S. NAVY CAG AIRCRAFT

C & M VOL. 16

PART 2 — ATTACK AIRCRAFT A-6 INTRUDER A-7 CORSAIR

Bert Kinzey & Ray Leader

THE COLORS & MARKINGS SERIES

The Colors & Markings Series has been designed to provide an ongoing affordable series of publications covering the paint schemes, squadron markings, special insignias, and nose art carried by many of the most important aircraft in aviation history.

Most books in this series will have sixty-four pages, and approximately one-quarter of the pictures will usually be in color. When older aircraft are presented, and extensive color coverage is not available, there will be less color, but more black and white photographs. Except for the front cover, no artist paintings will be used to show color, since these are notorious for having errors. When color photographs are not available, captions for the black and white photos will extensively describe the colors involved. We believe this will provide more accurate information than artists' renditions. But in most cases, there will be approximately 50 color photographs for a price under twelve dollars. Additionally, there will usually be at least one-hundred more black and white photographs. We will make it our policy to keep the photographs large so that the markings show clearly. Dates that the photographs were taken will often be included so that the reader will know the time frame for which the markings were appropriate.

Special volumes in this series will be released approximately once a year. These special volumes will be larger and will contain additional color.

The service life of each aircraft will dictate the specific format of the book. In some cases the book may cover a specific time frame or a given numbered air force. The active Air Force units that operated the aircraft may be the focus of one book, while Air National Guard units will be featured in another. In short, the format and scope of each book will be narrowed so that good coverage can be presented. However, it must be noted that it is impossible to provide definitive, much less complete, coverage of a given aircraft in a sixty-four page book. To do so is not our goal, but we do intend to provide as extensive coverage as possible in a continuing softbound series, and offered at a price even the reader on a budget can afford. In order to include as many photographs as possible, no extensive narrative will be included to cover the developmental or operational history of the aircraft. This is usually better provided in books designed primarily to present such information. This series will have only a brief introduction to explain the basic mission of the aircraft and the scope and format used for that particular book. It will be the captions for the photographs that will tell the story of the aircraft and its markings. To as great an extent as possible, photos that have not been published before will be used.

We have been fortunate to receive the cooperation and assistance from some of the best known aviation photographers in the world who have offered photos from their extensive collections for this series. With their help, we believe the Colors & Markings Series will be the best of its kind available anywhere.

THE AUTHORS

Bert Kinzey is an aviation writer who is best known for his "Detail & Scale" series which he created to cover the details of military aircraft. He has also written numerous magazine articles as well as manuals for the Department of the Army. He has served in the Army as an Air Defense Artillery Officer, during which time he commanded a Hawk guided missile battery in Korea, and later wrote and taught classes on airpower at the Army Air Defense School. As a civilian he served with the Army as a "subject matter expert" on the Soviet air threat, but now has turned his attention to full time writing. He is an avid aviation photographer and modeler, and is a member of the Aviation and Space Writers Association, the International Plastic Modelers Society, and the American Historical Society. He is also President of Detail & Scale, Inc.

Ray Leader is an Air Traffic Controller for the Federal Aviation Administration at the Atlanta International Airport. He began his aviation career when he entered the U.S. Army in 1958. Ray's interest in aviation led him into aviation photography which he has avidly pursued since 1962. He has one of the most extensive collections of aviation photographs in the world. He is a member of the International Plastic Modeler's Society, the American Aviation Historical Society, and the World Airline Historical Society. Ray operates "Flightleader Aviation Research and Photography," and travels extensively taking photographs for Detail & Scale and other aviation publications. He is the Vice President for Photography at Detail & Scale, Inc.

C & M VOL. 16
PART 2 — ATTACK AIRCRAFT
A-6 INTRUDER   A-7 CORSAIR

COLORS & MARKINGS
OF

U.S. NAVY CAG AIRCRAFT

Bert Kinzey & Ray Leader

TAB BOOKS
Blue Ridge Summit, PA

Airlife Publishing Ltd.
England

Copyright © 1990 BY DETAIL & SCALE, INC.

All rights reserved. No part of this publication may be reproduced in any form, stored in a retrieval system, or transmitted by any means, electronic, mechanical, or otherwise, except in a review, without the written consent of Detail & Scale, Inc.

This book is a product of Detail & Scale, Inc., which has sole responsibility for the contents and layout, except that all contributors are responsible for the security clearance and copyright release or all materials submitted. Published and distributed in the United States by TAB BOOKS, a division of McGraw-Hill, Inc., and in Great Britain and Europe by Airlife Publishing Ltd.

CONTRIBUTORS AND SOURCES:

Pete Wilson	Gary Minert	Naoki Nishimura	John Stewart
Don Logan	Lindsey Peacock	Jim Rotramel	Jean-Charles Boreux
Bob Stewart	Steve Miller	Mike Grove	Phillip Huston
Pete Berganini	Arnold Swanberg	Mike Campbell	Don Spering/A.I.R.
Bill Sides	Hideki Nagakubo	Jerry Geer	Flightleader
Brian Rogers	Bill Curry	Mick Roth	Military Aircraft Photographs
Steve Daniels	Bill Malerba	Peter Thompson	Centurion Enterprises

FIRST EDITION
FIRST PRINTING

Library of Congress Cataloging
in Publication Data

Kinzey, Bert.
  Colors & markings of U.S. Navy and USMC CAG aircraft.

  (C&M ; vol. 10-)
  (v.2- : Colors & markings ; v. 16)
  Spine title: U.S. Navy and USMC CAG aircraft.
  "A Detail & Scale aviation publication."
  Second ed. statement from label mounted on cover.
  Vol. 2 has title: Colors & markings of U.S. Navy CAG aircraft.
  Contents: pt. 1. Fighters — pt. 2. Attack aircraft.
  1. Airplanes, Military—United States—Identification marks. 2. Airplanes, Military—United States—Painting. 3. United States. Navy—Aviation. 4. United States. Marine Corps—Aviation. I. Leader Ray. II. Title: Colors & markings of U.S. Navy CAG aircraft. III. Colors and markings of U.S. Navy and USMC CAG aircraft. IV. U.S. Navy and USMC CAG aircraft. V. US Navy and USMC CAG aircraft. VI. Series. VII. Series: C & M ; vol. 10, etc.
  UG1243.K494 1988   359.9′4834   88-12181
  ISBN 0-8306-8534-0 (v. 1)
  ISBN 0-8306-7540-X (v. 2)

First published in Great Britain in 1990
by Airlife Publishing Ltd.
7 St. John's Hill, Shrewsbury, SY1 1JE

British Library Cataloging In
Publication Data

Kinzey, Bert, *1945 -*
  U.S. Navy C.A.G. Aircraft. - (Colors & marking series :
  v. 16)
  Pt. 2, Attack aircraft.
  1. United States Navy. Military aircraft
  I. Title II. Series
  358.4183

ISBN 1-85310-623-2

TAB BOOKS offers software for sale. For information and a catalog, please contact TAB Software Department, Blue Ridge Summit, PA 17294-0850.

Questions regarding the content of this book should be addressed to:

Reader Inquiry Branch
TAB BOOKS
Blue Ridge Summit, PA 17294-0214

Front cover: A-7E, 158018, was the CAG aircraft for VA-22 when this photograph was taken in 1972. VA-22 was part of the air wing aboard the USS CORAL SEA at that time. (Nagakubo)

Rear cover, top: The CAG aircraft from VA-97 was A-7E, 156872, when this photograph was taken at NAS Lemoore on May 6, 1979. In the background is the CAG aircraft from VA-22. (Rhodes via Spering/A.I.R.)

Rear cover, bottom: This A-6E, 155704, carried VA-115's CAG markings on May 28, 1976. The photograph was taken at NAF Atsugi, Japan. (Flightleader Collection)

INTRODUCTION

A-6E, 152950, was assigned to VA-165. The CAG markings on the tail of this aircraft were among the most colorful used by Intruder squadrons. (Flightleader Collection)

Volume 10 in Detail & Scale's Colors & Markings Series was about U.S. Navy and Marine Corps fighter aircraft that had been painted in special markings for the commanding officers of carrier air groups and Marine air wings. In that book we included the F-8 Crusader, the F-4 Phantom, and the F-14 Tomcat. In the years during and since the war in Vietnam, it is these three fighters that have been assigned to the various fighter squadrons in both the Navy and the Marine Corps.

This new title in the Colors & Markings Series continues our coverage of U.S. Navy CAG markings by illustrating the two primary attack aircraft that have been flown by the Navy's attack squadrons since the end of the war in Vietnam. These are the A-6 Intruder and the A-7 Corsair II, and they are presented in that order in this publication.

While both of these aircraft carry out the ground attack mission, they are designed to perform this mission in different ways. The A-6 is a two place, highly sophisticated all-weather aircraft that can attack targets in poor weather and at night. Intruder flight crews will tell you that they prefer to fly when there is "dirty" weather to and from the target. Today, the A-6 has passed its silver anniversary of operational service, and remains the Navy's and the Marine's primary attack aircraft. With no replacement yet past the planning stage, the A-6 will be in the operational inventory for several years to come, and will probably serve into the twenty-first century.

The A-7 Corsair, particularly the A-7E, is also sophisticated in that it has avionics which enable it to deliver a wide variety of ordnance against ground targets with a great deal of accuracy. However, it performs its mission in clear weather, since it does not have the all-weather capability of the A-6. Having served throughout most of the war in Vietnam and during the years since its conclusion, the A-7 is now being replaced with the F/A-18 Hornet. As this is written, the Corsair II remains in a few active fleet squadrons as well as reserve squadrons, but its days of operational fleet service are almost at an end. It has performed very well, and the Air Force's A-7D and A-7K variants will remain in Air National Guard service for some time to come.

As we explained in our earlier book on CAG markings, Navy squadrons usually designate one of their aircraft for the CAG or **C**ommander of the **A**ir **G**roup to fly. However, in practice, these aircraft are flown routinely by any pilot in the squadrons. The three-digit nose number on these aircraft, called a modex, begins with the squadron's number in the air wing and ends with 00. The last two digits are usually on the tail as well, and the two zeros are often referred to as "double nuts." On some CAG aircraft, these two zeros are painted in the form of mechanical nuts as will be illustrated in this book. For many years, and to a limited extent today, CAG aircraft were painted in special, and usually more colorful markings than the rest of the aircraft in the squadron.

When both the A-6 and A-7 entered operational service, the light gull gray (FS 16440) over white paint scheme was official for these two aircraft. During the time that this paint scheme was authorized, squadron markings, and especially CAG markings, were far more colorful than they are today. However, it is clearly illustrated in this book that, as a general rule, the Corsairs were considerably more colorful than the Intruders. While the A-6 squadrons did use a fair amount of color in their markings, it seldom was very extensive, and the CAG markings did not markedly differ from those used on other aircraft in the squadron. CAG markings that did use major splashes of color were usually seen earlier in the Intruder's service, and they became more and more subtle as time went on. There were exceptions to this rule, but they were relatively rare. A-7s, on the other hand, often had their entire vertical tails painted in squadron colors, and CAG markings were both large and elaborate on many aircraft. But much has changed, and on today's tactical schemes, the commander's name and the acronym "CAG" may be the only special markings on the aircraft. Some aircraft carry only the 00 modex with no CAG markings at all.

In 1976, the United States celebrated its bi-centennial anniversary, and many military aircraft received special red, white, and blue markings to commemorate the event. These markings varied from rather small insignias to examples where the entire aircraft was repainted. CAG aircraft, which already had unusually colorful markings, became even more colorful when the bi-centennial markings were added. As a result, some of the most colorful aircraft in the history of the U.S. Navy could be seen in the skies, on the carriers, and at naval air stations during 1976. A number of these bi-centennial CAG aircraft are illustrated on the pages that follow.

But the bi-centennial year was hardly over when the Aircraft Combat Survivability Branch directed that combat aircraft in the U.S. Navy be painted in more subdued or low visibility markings. This was to reduce both their visual and infrared signatures. Colorful markings began to be replaced with blacks and grays, and eventually, the light gull gray over white paint scheme was replaced with the present tactical scheme. When this happened, color almost completely disappeared. Today, some color is reappearing on F-14 Tomcats, and to a lesser extent on F/A-18 Hornets. But the subsonic A-6 Intruders and A-7 Corsairs remain in their colorless tactical schemes, and it seems unlikely that much if any color will ever be seen again on these aircraft.

In this publication we have placed our emphasis on the older, more colorful schemes and markings. However, we have tried to show examples of how these markings have changed over the years, and the low visibility and tactical schemes are also included. As is the case with almost all of the books in this series, we have arranged the coverage of these aircraft on a unit-by-unit basis. At least one example of a CAG aircraft is shown from each of the fleet and reserve squadrons that have flown these two aircraft. The squadrons are presented in numerical order. Readers should remember that squadrons assigned to the Atlantic Fleet have tail codes starting with the letter **A,** while those assigned to the Pacific Fleet have tail codes that begin with **N.** At times, Pacific Fleet squadrons have made deployments to the Atlantic or Mediterranean Sea, and have temporarily carried tail codes beginning with an **A.** Likewise, Atlantic Fleet squadrons have deployed to the Pacific, and have carried tail codes that start with an **N.** This was particularly true during the war in Vietnam. Several of these "opposite ocean" deployments and changes in tail codes are pointed out in this publication. In all, there are sixteen A-6 and thirty-six A-7 squadrons included. This includes the single A-7 from VA-125 that is illustrated below. We have made an attempt to show more than one example of CAG markings for most units, and have illustrated both dramatic and subtle changes in markings.

On the pages that follow is the most extensive coverage of CAG markings applied to the Intruder and the Corsair II ever included in a single publication. Certainly there have been other CAG markings carried on these two aircraft than those illustrated in this book, but every attempt has been made to show as many CAG markings as possible in the space that is available. It is our hope that this book will provide an historical record of these markings for aviation enthusiasts, since no Intruder or Corsair II is likely to ever carry the colorful CAG markings again. We also hope that modelers will use this book as a reference to accurately mark replicas of some of the aircraft that are illustrated. Such models would be among the most colorful in any collection.

Many contributors have provided numerous photographs for this book in order to make it as complete as possible. Their names are listed on page 2, and the authors express a sincere word of appreciation to all of them.

*A-7B, 154436, was one of the few aircraft assigned to VA-125 that ever carried CAG markings. The color of the large lightning bolt on the rudder and tail is unknown. The **500** modex, painted on both the fin cap and the nose of the aircraft, was flat black.*
(Flightleader Collection)

A-6 INTRUDERS
VA-34 BLUE BLASTERS

A-6E, 158043, was assigned to the USS JOHN F. KENNEDY and VA-34 when it was photographed at NAS Oceana, Virginia, on April 26, 1974. The squadron emblem was painted within the blue diamond on the tail of the aircraft. Behind the skull was a fan in the CAG colors (top to bottom) green, red, yellow, blue, brown, yellow, and black. The black **AB** tail code was shadowed in white.
(Spering/A.I.R.)

The right side of the same CAG aircraft is seen in this photograph. It was taken in September 1976, when the aircraft had VA-34's bi-centennial markings on the upper portion of the fuselage behind the canopy. These markings consisted of red, white, and blue stars on a gold circle and the words, **"SPIRIT OF 76,"** painted in red, white, and blue.
(Spering/A.I.R.)

When VA-34 started using the low-visibility markings in the 1980s, their CAG aircraft also reflected this change. A-6E, 161231, was photographed while taxiing out for a training mission at NAS Fallon, Nevada. Notice the small squadron emblem painted in gray on the white rudder. The **AB** tail code was positioned horizontally across the tail in a contrasting darker gray. The modex on the nose and fin cap were still lettered in black.
(Grove)

VA-35 BLACK PANTHERS

*Above: A-6A, 154125, which was assigned to VA-35, was photographed at NAS Oceana, Virginia, in May 1969. The aircraft had basic CAG markings with the **500** modex on the nose and fin cap. Notice that the Intruder still had Pacific Fleet **NG** tail codes from VA-35's deployment aboard USS ENTERPRISE for a tour off the coast of South Vietnam.* (Spering/A.I.R.)

Right: This photograph is a close-up view of the bomb mission markings on the aircraft seen above. (Spering/A.I.R.)

*A-6E, 149956, displays VA-35's subtle bi-centennial markings in this photograph which was taken in September 1976. The aircraft had dark green **AJ** tail codes on a white circle which was edged in dark green. The black panther was painted between the tail codes. As a bi-centennial marking, **USS NIMITZ** was lettered in red, white, and blue above **NAVY** on the fuselage.* (Spering/A.I.R.)

*VA-35 had changed to the tactical paint scheme by the time this photograph of A-6E, 161688, was taken on April 19, 1986. The markings were a contrasting gray, except for the black **500** modex on the nose. Notice the placement of the panther on the rudder and the horizontal arrangement of the **AJ** tail code.* (Wilson)

VA-42 GREEN PAWNS

VA-42 usually did not paint CAG markings on any of their Intruders, however, this aircraft is an exception. A-6A, 152935, was photographed at NAS Moffett, California, in August 1969, and had **CAPT JIM MEGIN CAG-4** lettered on the canopy rail. It also had the **500** modex on the nose and fin cap to denote the CAG aircraft. (Spering/A.I.R.)

This photograph of 160421 was taken while the aircraft was at NAS Oceana on September 5, 1977. Notice the addition of the large green pawn on the tail. The **AD** tail code was black and edged with white, and the name, **CAPT. T.E. SHANAHAN,** was stencilled in white on a green rectangle under the canopy. (Miller)

A-6E, 158531, had slightly larger **AD** tail codes than the aircraft seen above. **CAPT G.M. HESSE MATWING ONE** was lettered in white on a green rectangle under the canopy. (Spering/A.I.R.)

VA-42 had changed to low visibility markings by the early 1980s. A-6E, 149955, is illustrated here in a subdued gray paint scheme with contrasting gray markings. Notice the smaller pawn on the white rudder. (Grove)

VA-52 KNIGHT RIDERS

A-6E, 155670, was assigned to VA-52, when it was photographed at Nellis AFB, Nevada, in February 1977. The only CAG marking for this aircraft was the black **500** modex on the nose and at the top of the rudder. The knight's head at the top of the tail was blue, and the **NH** tail code was black. The carrier name, **USS KITTY HAWK,** was located above **NAVY** on the fuselage.
(Logan)

A change in tail codes for VA-52 had occurred by the time this photograph of A-6E, 155637, was taken at NAS Fallon, Nevada, in 1980. The **NL** on the tail and the **500** modex on the nose were painted black and shadowed with white. Notice the addition of the colorful plumes on the knight's head for the special CAG markings. A large blue lance had been added along the side of the fuselage. The name on the canopy rail was **C.A.G. DAVE EDWARDS.** (Grove)

This A-6E TRAM was painted in the tactical scheme when it was photographed at Eglin AFB, Florida, in September 1989. The carrier name, **USS CARL VINSON,** was painted on the fuselage. All markings were a contrasting gray except for the **500** modex on the nose and top of rudder, and the **VA52** on the wing tank. These were flat black. (Swanberg)

VA-55 SEA HORSES

One of the CAG Intruders used by VA-55, was A-6E, 161675, which was photographed at NAS Oceana on May 5, 1984. The aircraft was painted in the tactical scheme with contrasting gray markings. The **AK** tail code was painted horizontally on the tail, and **USS CORAL SEA** was lettered in a dark gray above **NAVY** on the fuselage. These markings are so faint that they are impossible to see in this photograph. (Campbell)

Above: The CAG aircraft for VA-55 had been changed to A-6E, 158051, when this photograph was taken in October 1988. The squadron emblem was painted in black on the rudder. (Flightleader)

Right: This close-up of the squadron markings on the tail shows the addition of color for the CAG aircraft. (Flightleader)

VA-65 TIGERS

*Above: One of the more colorful Intruder squadrons was VA-65, as evidenced by this photograph of A-6E, 158531, taken on April 26, 1974. The bright orange stripes on the fuselage and tail were edged in black, as was the **AG** tail code.* (Spering/A.I.R.)

Left: This close-up reveals details of the special CAG markings on the rudder of the same aircraft shown above. (Spering/A.I.R.)

The left side of the same CAG aircraft is shown here to illustrate the markings painted on that side of the aircraft. (Spering/A.I.R.)

A-6E, 161095, shows the markings that VA-65 used as they started their transition to low visibility markings. All markings are flat black on the gray over white paint scheme. (Wilson)

*By the time A-6E, 151804, was photographed on May 11, 1984, VA-65 was using the tactical scheme on most of their aircraft. All markings were in a contrasting gray, including the carrier name, **USS DWIGHT D. EISENHOWER,** on the fuselage. Again, on this tactical scheme, the markings are almost impossible to see.* (Kinzey)

*A-6E, 155599, was the CAG aircraft for VA-65 in September 1985. The squadron emblem was painted on the rudder in a dark gray, and **CDR MIKE MATTON CAG** was on the canopy rail.* (Flightleader)

VA-75 SUNDAY PUNCHERS

VA-75, is known as the Sunday Punchers. Their CAG aircraft in the early 1970s was A-6E, 155631, which is illustrated here. The horizontal colored stripes on the rudder represented the CAG's colors, and the **AC** tail code was assigned to the air wing aboard the USS SARATOGA. (Nagakubo)

By 1973, VA-75 had changed their markings to the blue and yellow flashes and stars seen on the white tail of this aircraft. The **AC** tail code was black using oriental style lettering. (Spering/A.I.R.)

This right rear view of A-6A, 152587, gives a better view of the markings painted on the aircraft. (Geer)

*VA-75 was flying A-6Es when this photograph was taken on the deck of the USS SARATOGA. **CAG CDR. LOU SCHRIEFER** was lettered under the canopy rail.* (Spering/A.I.R.)

Left: This close-up of the tail of 158790 illustrates the new CAG markings that were painted on the rudder. The colored stars surrounding the squadron badge were (clockwise from top) yellow, red, green, black, and orange. (Flightleader)

*This photograph of A-6E, 158051, was taken in February 1978. The name under the canopy rail was **CAG KOBER**.* (Grove)

VA-75 also used the tactical paint scheme as seen in this photograph of A-6E, 160997, taken at NAS Oceana in May 1986. All markings, except for the black modex, were in contrasting shades of gray. (Paul via Geer)

VA-85 BLACK FALCONS

Above: A-6E, 158792, was the CAG aircraft for VA-85 in September 1974. The unit was assigned to the USS FORRESTAL and had special CAG markings painted on the rudder. (Spering/A.I.R.)

Left: This close-up shows the details of the special markings on the rudder of A-6E, 158792. The colors in the markings were (top to bottom) red, yellow, blue, orange, green, and black. (Spering/A.I.R.)

Another variation of CAG markings is seen in this photograph of A-6E, 158792. The horizontal band had been eliminated from the tail, and the special markings on the rudder had been changed to horizontal bars, using the same colors as before. (Spering/A.I.R.)

By July 1977, another change to VA-85's CAG markings had taken place. The entire rudder was painted with horizontal bars. Notice that the squadron emblem, which was located on the tail, had red and white feathers added to it. **CDR FRED METZ COMMANDER CVW-17** *was painted in black below the canopy.* (Stewart)

This left side view of the same aircraft seen above shows a change to the squadron emblem on the tail. The red and white feathers had been eliminated when this photograph was taken in August 1977. (Swanberg via Geer)

At left is an overall view of A-6E, 158528, which was the CAG aircraft for VA-85 in August 1980. **CAG KING BRENNER** *was lettered in black under the canopy. At right is a close-up view of the different CAG markings on the rudder.* (Both Grove)

VA-95 SKY KNIGHTS

A-6A, 154144, was the CAG aircraft for VA-95 when this photograph was taken at NAS Whidbey Island, Washington, on February 6, 1974. Special CAG markings were painted on the aircraft's rudder. (Remington via Spering/A.I.R.)

A change in the CAG markings had taken place by the time A-6A, 156995, was photographed at NAS Miramar, California, in November 1975. The shape and color of the trident on the tail had changed, and the CAG markings on the rudder were small horizontal stripes. A dark stripe on the fin cap contained the **00** modex indicating the CAG's aircraft. (Geer)

VA-115 CHARGERS

The CAG aircraft from VA-115 was photographed as it taxied out for a mission at NAF Atsugi, Japan. The special CAG markings were confined to the fin cap on this aircraft.   (Wada via Spering/A.I.R.)

The right side of the aircraft seen above was photographed in September 1974. The squadron badge was located on the fuselage behind the canopy.   (Taylor via Geer)

A more conservative paint scheme for the VA-115's CAG aircraft was used on A-6E, 152610. This photograph is dated September 15, 1977. A yellow chevron was painted on the front of the tail, and a white arc and colored spikes were added as the CAG markings. The fin cap was dark green, and had a row of yellow stars painted within it. A yellow stripe bordered the green fin cap at the bottom. **CAG CDR KIRK CONNELL** was lettered in yellow on a green banner under the canopy.
(Spering/A.I.R.)

VA-128 GOLDEN INTRUDERS

The only marking that indicated that A-6A, 155644, was the CAG aircraft for VA-128 was the **800** modex on the nose. This photograph was taken at Edwards AFB, California, in May 1969. The squadron badge was located on the fuselage behind the canopy. (Spering/A.I.R.)

A change in the markings from those illustrated on the aircraft above can be seen in this photograph dated May 18, 1979. The tail markings and the fin cap were gold and white. The name, **CITY OF OAK HARBOR,** was in black above **NAVY** on the fuselage, and is barely discernable in this photograph. (Tunney via Spering/A.I.R.)

VA-145 SWORDSMEN

Very attractive markings were applied to the tail of A-6E, 155708. The large sword on the tail contained the various colors for CAG markings.
(Grove)

At left is an overall view of A-6E, 155648, which was the CAG aircraft for VA-145 in 1981. The sword on the tail was black. At right is a close-up view of the left side of the tail which illustrates the CAG markings in better detail. They consist of small colored stars painted on the fin cap. The colors of the stars were (left to right) green, black, blue, red, black, and orange.
(Both Grove)

Quite a few changes had taken place by August 4, 1984, when this photograph was taken. The **NE** tail code was still used, but the carrier assignment had been changed to the USS KITTY HAWK. The markings were toned down and smaller than those used previously. **CAPT ERNIE CHRISTENSEN** was lettered in black under the canopy. (Remington via Spering/A.I.R.)

VA-155 SILVER FOXES

VA-155 was established as an Intruder squadron on September 1, 1987, but was short lived. This is the only photograph of a CAG Intruder from VA-155 included in this book, and it illustrates the only A-6 known to have been painted in CAG markings for this squadron. A-6E, 152930, is shown here as it appeared at Pasco, Washington, in July 1988. The aircraft was painted in the tactical scheme with contrasting gray markings, except for the black **500** modex on the nose and fin cap. **CAPT AUSTIN CHAPMAN CVW-17** was painted in black under the canopy. Note that VA-155 was formerly an A-7 squadron as illustrated on page 52. It was disestablished as an A-7 squadron in late 1977 along with the decommissioning of the USS FRANKLIN D. ROOSEVELT. (Swanberg)

VA-165 BOOMERS

At left is A-6E, 152950, the CAG aircraft for VA-165, as it appeared on September 22, 1975. At right is a close-up of the special markings on the tail of this aircraft. The colors in the chevron were (front to back) black, yellow, brown, blue, red, and green. (Both Curry)

A different CAG aircraft is seen in this 1978 photograph of A-6E, 155660. The carrier name, **USS CONSTELLATION,** was painted in black above **NAVY** on the fuselage. (Roth)

VA-176 THUNDERBOLTS

A-6A, 151821, was the CAG aircraft for VA-176 when it was photographed in the early 1970s. The carrier name, **USS ROOSEVELT,** was painted in black on the fuselage. The **AE** tail codes were black, shadowed with white, and had an orange lightning bolt between them. Colored stars were painted along the back edge of the rudder for the CAG markings. (Flightleader)

A different Intruder was being used as the CAG aircraft for VA-176 by the time this photograph was taken in June 1974. The **AE** tail codes were painted white and outlined in black. A white fist held the orange lightning bolt. The colored stars along the bottom of the rudder were (left to right) black, blue, green, orange, yellow, and red. (Geer)

VA-176 was using the tactical scheme by the time A-6E, 159317, was photographed in late 1989. The aircraft had contrasting gray markings, except for the black **AE** tail code and the **500** modex painted on the nose and fin cap. **USS FORRESTAL** and **CAPT PAT HAUERT CAG** were also painted in black. (Swanberg)

VA-196 MAIN BATTERY

A-6A, 155600, was the Intruder used by VA-196 as their CAG aircraft when this photograph was taken on February 17, 1976. The aircraft displayed a diagonal row of colored spades down the rudder as its special CAG markings.
(Bergagnini)

The left side of A-6E, 159900, was photographed on December 3, 1977. Notice that the size of the colored spades on the rudder was smaller than those seen in the photograph above.
(Logan)

The CAG aircraft used by VA-196 during the bi-centennial celebration was A-6E, 155597. A bi-centennial marking, in the form of a red and blue **76**, was located on the bottom of the white rudder.
(Flightleader)

*A-6E, 154170, was the CAG aircraft being used by VA-196 in November 1979. It further illustrates the markings used by that squadron. **USS CORAL SEA** and **CVW 14** were lettered in black above and below **NAVY** on the fuselage.* (Grove)

*At left is a close-up view of the special CAG markings painted on the tail of the aircraft seen above. At right is a slight variation of the CAG markings painted on the same aircraft in 1981. The orange horizontal stripe had disappeared from the tail, and was replaced by the black and orange flash across the fin and rudder. An orange circle, which was edged in black, contained a large black spade and white **NK** tail codes. The colored spades on the rudder appear to be larger than those seen in the photo at left.* (Both Grove)

*This photograph of A-6E, 159570, was taken as the aircraft taxied out for a mission in October 1982. The horizontal orange stripe was on the tail again. The band was extended across the top of the rudder, and consisted of the colors (left to right) black, blue, green, red, and yellow. These were the CAG colors for this aircraft. A small black bat was painted at the bottom of the rudder, and **CDR R. CASH CAG CVW 14** was lettered in black on an orange rectangle below the canopy.* (Grove)

A-7 CORSAIR IIs
VA-12 CLINCHERS

A-7E, 157563, was VA-12's CAG aircraft when this photograph was taken on June 22, 1972. The entire rudder was painted with colorful markings for the CAG. (Malerba)

The colorful CAG markings were not present on the rudder of Corsair II, 157472, when it was photographed on April 12, 1975. **CDR JIM FLATLEY CVW-7** was painted under the canopy. (Flightleader)

VA-12 transitioned to the tactical scheme, and their new markings were illustrated in this view of A-7E, 157452. All markings are a contrasting gray, including the carrier name, **USS DWIGHT D. EISENHOWER,** painted above **NAVY** on the fuselage. Notice that **VA-12** has been relocated to the tail cone instead of being on the front of the tail as seen previously. (Thompson)

VA-15 VALIONS

A-7B, 154492, was the aircraft that carried the CAG markings for VA-15 during 1970. Notice the vertical row of colorful stars painted on the rudder. The front of the tail was painted light blue, and all other markings were black. (MAP)

A change in the squadron's tail markings had taken place by the time this photograph was taken in December 1975. The tail was painted white, and a light blue stripe, edged in yellow, was painted along the leading edge of the fin and up to the top of the rudder. The colorful stars were still on the rudder to represent the CAG markings. The lion on the tail and the **AE** tail code were painted blue, and edged in yellow. The carrier name, **USS AMERICA,** was in black above **NAVY.** (Sides)

A-7E, 159655, was the CAG aircraft being used by VA-15 when it was photographed at NAS Oceana on October 20, 1977. The colorful stars on the rudder had been replaced with blue ones that were edged in yellow.

(Spering/A.I.R.)

VA-22 FIGHTING REDCOCKS

A-7E, 159980, was the CAG aircraft for VA-22 when this photograph was taken on October 29, 1977. The colorful horizontal stripes painted on the rudder, and the unusual double nuts on the fin cap comprised the special markings for the CAG. **CDR SNUFFY SMITH CAG** *was painted in white on the blue rectangle beneath the canopy. (Bergagnini)*

A different aircraft is illustrated in this right side view of 157466, however, the markings are the same as seen in the photograph above. (MAP)

Another one of VA-22's Corsair IIs was painted in CAG markings by the time this photograph was taken in May 1978. No carrier name appeared on the aircraft at that time. (Swanberg via Geer)

Numerous changes in CAG markings had taken place by the time this aircraft was photographed at NAS Lemoore, California, in September 1980. The colorful bars on the rudder had been replaced with blue bars. A **15** was painted in black in the middle of the rudder, and was surrounded with small colored stars. **CAG** was lettered in black on the fin cap, and **CAG CAPT DAVE EDWARDS COMCARAIRWING 15** was in blue under the canopy. (Bergagnini)

Additional changes are evident in this 1982 photograph of A-7E, 159998, taken while the aircraft taxied out for a mission. It was painted overall gloss gray, a scheme that was relatively rare on A-7s, and the tail code had been changed to **NH**. **USS ENTERPRISE** was stencilled in black on the tail cone, and the number **11** on the rudder was surrounded with small colorful stars. **CDR F. LEE TILLOTSON COMCARAIRWING 11** was painted in blue under the canopy. (Grove)

When VA-22 changed to the tactical scheme, all markings, including the **CVW 11** and the emblem on the rudder, were painted in a light contrasting gray. (J. Stewart)

VA-25 FIST OF THE FLEET

The CAG aircraft used by VA-25 in 1970 was A-7E, 157483. The entire tail was painted dark green, and had the **NE** tail code positioned on a yellow disc. The red lightning bolt was edged in yellow. Colorful diamonds were painted on the rudder to indicate this as being the CAG aircraft.
(Flightleader)

A minor change in markings had taken place by the time this photograph of A-7E, 157451, was taken in April 1972. A large pair of nuts had been painted in green on the fin cap and nose to represent the 00 modex, and the colorful diamonds on the rudder had disappeared. (Flightleader)

Even more changes in markings are seen in this photograph of A-7E, 158666, which was taken at Dobbins AFB, Georgia, in 1978. The dark green began at the radome, ran along the canopy and the spine of the fuselage, and extended all the way back to the tail. The **NE** tail code was positioned vertically on the rudder in black. The squadron emblem was painted inside the yellow disc on the tail. Note the special colorful markings at the base of the rudder and just aft of the green area behind the canopy.
(Curry)

*A move toward low-visibility markings can be seen in this view of A-7E, 158666. The green tail had been replaced with a dark green stripe painted horizontally from the spine of the fuselage to the bottom of the rudder. The words, **FIST OF THE FLEET**, were in white on the green stripe. The squadron markings on the tail were smaller than previously seen. (Grove)*

*Right: This is a close-up of the special CAG markings on the right side of the tail. The **NE** tail code was green, shadowed with yellow, and colorful CAG markings surrounded the 2 at the bottom of the rudder. (Grove)*

The markings on the left side of the tail of the same aircraft are seen in this view. The squadron markings were still in full color. (Grove)

VA-27 ROYAL MACES

A-7E, 158832, was VA-27's CAG aircraft in 1973, when this photograph was taken as it taxied back to the ramp. The mace on the tail was black with a red handle. Colorful spikes served as the CAG's markings. The **NK** tail code and **400** modex on the nose were black, shadowed with white. The chevrons on the white rudder were green.
(Remington via Spering/A.I.R.)

This view of A-7E, 157536, shows the aircraft taxiing out for a training mission. Notice that the handle of the mace has been changed from red to black, and the chevrons on the rudder are painted in various colors for the CAG. Gold double nuts are on the fin cap.
(Grove)

A slight change had taken place by the time this photograph was taken in 1981. The mace on the tail was black overall, as was the **NK** tail code. The dark green fin cap contained **00** in white. The carrier name, **USS CORAL SEA,** on the tail cone indicates a change in assignments from that seen in previous photographs. (Grove)

VA-27 had adopted the tactical scheme by the time this photograph was taken in late 1984. Several changes in markings appear here, and include the fist holding the mace on the tail. **VA-27 USS CARL VINSON** was in black on the bottom of the tail. A white **400** modex and a black **CAG** were also painted on the aircraft.
(Grove)

VA-37 BULLS

A-7A, 153161, was VA-37's CAG aircraft when this photograph was taken. The fin cap was light blue with a white **00**. The colorful squadron emblem was located below the black **NH** tail code. The **NH** tail code indicates a Pacific Fleet deployment for this Atlantic coast squadron.

(Flightleader)

The same aircraft shown above was photographed at NAS Cecil Field, Florida, in November 1970, after it was reassigned to the Atlantic Fleet. The carrier name, **USS SARATOGA**, was painted in black above **NAVY**. Notice the addition of CAG markings in the form of colorful diamonds on the rudder, and **CVW3** arranged vertically in front of the rudder.

(Spering/A.I.R.)

This photograph of A-7E, 158826, was taken while the aircraft was at Nellis AFB, Nevada, in 1977. The **AC** tail code was black, and was shadowed in white. Colorful lightning bolts were painted on the white rudder, and **AKRON 37 BULLS** was on the bottom of the tail under the light blue horns. **CVW-3** was in white on the light blue fin cap.

(Malerba)

A beautiful addition to the tail of A-7E, 157559, was the large Schlitz Malt Liquor bull painted in black and white. Smaller colored lightning bolts were on the rudder, and the carrier name, **USS SARATOGA**, was lettered in black on the fuselage.

(Curry via Spering/A.I.R.)

Several subtle changes in the markings for VA-37's CAG aircraft are evident in this photo of 158830. Notice that the bull was painted black and blue, and different CAG markings were painted on the rudder. (Grove)

Another variation to VA-37's CAG markings is illustrated on this aircraft, which was photographed while taxiing out for a mission. Small colorful bulls had been painted on the white rudder. (Grove)

*VA-37 had switched to the tactical scheme by the time this photograph was taken in 1985. The unit had been assigned to the USS FORRESTAL and had **AE** tail codes painted on the aircraft. The size of the bull on the tail was smaller than the ones used during the colorful time period.* (Grove)

*Another change in VA-37's CAG markings is evident in this 1989 photograph, taken on the flight deck of the USS FORRESTAL. The squadron badge was located inside an inverted triangle with **CVW SIX** and the bull painted in black. The **VA37** was white.* (Kinzey)

VA-46 CLANSMEN

A-7B, 154462, was VA-46's CAG aircraft when this photograph was taken at NAS Cecil Field on October 4, 1971. The Clansmen used a plaid band painted around the forward fuselage and horizontally across the tail. The colorful horizontal bars on the rudder were the special markings for the CAG.   (Strandberg via Spering/A.I.R.)

At left is a view of VA-46's CAG aircraft, which was painted in attractive colors during the bi-centennial celebration. The fin cap was painted in plaid similar to the band at the bottom of the tail. **CLANSMEN** was stencilled in black, and was edged with white, along the bottom of the tail. At right is a close-up view of the red, white, and blue bi-centennial streamer that was positioned above the intake.   (Both Sides)

A-7B, 154462, was the CAG aircraft when it was photographed on August 13, 1977. The bi-centennial markings had disappeared, and a smaller **AB** tail code was painted in black at the top of the tail. **CAPT. JACK PRESLEY CAG** was in white on a blue rectangle under the canopy.   (Rogers)

VA-56 CHAMPIONS

A-7A, 154547, is shown here as it appeared on October 17, 1970. This Corsair II had bands of unusual colors on the rudder. They were (top to bottom) red/yellow, gray/orange, dark green/black, and navy/purple. The **AH** tail code and the **400** modex were black, shadowed with red. *(Roos via Spering/A.I.R.)*

A-7A, 152675, was photographed in the landing pattern at NAF Atsugi, Japan, on January 7, 1975. The special CAG markings on this aircraft consisted of horizontal stripes painted on the fin cap. *(Nishimura)*

VA-66 WALDOMEN

The Corsair II from VA-66 that carried CAG markings on May 16, 1971, was A-7E, 157544. The yellow and blue stripes, painted diagonally across the tail, had a **CVW 7** badge painted on them. Notice the unusual addition of the **C** above the **AG** tail code to spell out "CAG." (Spering/A.I.R.)

A change in the markings for VA-66 had taken place by 1978. The stripes on the tail were different than those seen in the photograph above, but were still yellow and blue. The **AG** tail code was painted using open style black letters on the yellow circle, and had blue shadowing. The carrier name, **USS DWIGHT D. EISENHOWER,** and **VA66** were black. (Grove)

Another revision to the markings is seen on A-7E, 160563, which was photographed in May 1981. A dark blue triangle, edged in red, and containing **CVW 7** in red, was painted on the yellow circle on the tail. The **AG** tail code was black and was split by the triangle. **CITY OF WALDO** was lettered in blue on the fuselage behind the canopy. (Grove)

VA-66 had started using the tactical scheme by the time this photograph was taken in 1984. The markings were painted in a contrasting gray except for the black **300** modex on the nose and fin cap. (Boreux)

VA-72 BLUE HAWKS

This colorful Corsair II was the CAG aircraft being used by VA-72 in late 1974. The colorful checkerboard pattern on the rudder was for the CAG, and made a nice addition to the aircraft. (Sides)

During the bi-centennial celebration, VA-72 added a red, white, and blue stripe around the front of the fuselage. The years **1776 - 1976** were lettered in black on the white portion of the stripe, but they are too small to be visible in this photograph. (Sides)

A-7E, 160552, was photographed on the flight deck of the USS AMERICA on October 20, 1982. **VA-72** was in black on the upper fuselage behind the canopy. (Daniels)

VA-81 SUNLINERS

*A-7E, 158028, was photographed at NAS Cecil Field, Florida, in September 1975. The **AA** tail code and **400** modex on the nose were black with red shadowing. The large red chevron on the tail was edged with black. The small chevrons, painted in various colors on the rudder, were the special CAG markings.* (Buchanan via Geer)

*Even though VA-81 was using the low-visibility paint scheme and markings when this photograph was taken, this CAG aircraft was still attractive. The white markings painted over the gray scheme really stood out. The lightning bolt that passed through the **AA** tail code was red.*
(Pocock via Spering/A.I.R.)

*By the time this 1982 photograph of Corsair II, 160719, was taken, VA-81 was back to more colorful markings for their CAG aircraft. The colorful chevrons and **CVW 17** painted on the rudder were a nice touch.* (Grove)

VA-82 MARAUDERS

The CAG aircraft from VA-82 in 1971 was A-7E, 157565. The markings on the tail were light blue and white, and the stripe behind the cockpit was light blue. (Strandberg via Malerba)

A-7E, 159289, was photographed while awaiting its next flight. Notice the colored ribbon held in the eagle's beak on the tail. The carrier name, **USS NIMITZ,** was in black on the tail cone, and **CDR JUDD SPRINGER CAG** was in black under the canopy. (Geer Collection)

All of the markings on the tail of 158827 were light blue, but they were reduced in size from those seen previously. (Stewart)

VA-83 RAMPAGERS

The CAG aircraft for VA-83 was A-7E, 157459, when this photograph was taken in August 1971. The aircraft had colorful horizontal bars on the rudder below the **AA** tail code, and the large ram's head on the tail had red eyes. The fuselage band was painted blue, and was edged in yellow.
(Spering/A.I.R.)

The same markings were also used on a different CAG aircraft as illustrated in this 1977 photograph. **CDR MOOSE MYERS** was painted in red and blue on a white ribbon under the canopy. The ribbon was a small representation of bi-centennial markings.
(Flightleader)

More subdued markings had made their appearance with VA-83 by the time this photograph was taken in late 1978. **CDR BUD LINEBURGER CVW-17** was lettered in white on a blue rectangle beneath the cockpit. (Swanberg)

VA-86 SIDEWINDERS

A-7A, 153136, was VA-86's CAG aircraft when this photograph was taken in 1969. The aircraft had **NL** tail codes, and the carrier name, **USS CORAL SEA,** was on the tail cone. Both of these markings indicated a tour in the Pacific. The orange, black, and white rattlesnake on the tail was very striking (no pun intended), and the colored diamonds on the rudder were markings for the CAG. (MAP)

Larger colored diamonds than those seen in the photo above were painted on the rudder of this aircraft. **SIDEWINDERS** was lettered in white on an orange band across the bottom of the tail. The **AJ** tail code and carrier name, **USS AMERICA,** provide evidence that the unit had been reassigned to the Atlantic Fleet. (Spering/A.I.R.)

VA-86 had transitioned to the A-7E by the time this photograph was taken in 1975. All of the diamonds on the rudder were orange. (Spering/A.I.R.)

During the bi-centennial celebration, VA-86 added the banner with **DONT TREAD ON ME** under the rattlesnake on the tail. Notice the addition of the large orange diamonds painted on the top of the fuselage behind the cockpit. (Flightleader)

VA-87 GOLDEN WARRIORS

*A-7E, 159968, was the aircraft being used by VA-87 for the CAG when this photograph was taken in July 1978. The Indian head on the tail was black with a yellow headband. A vertical row of colored stars was painted on the rudder for CAG markings. The **AE** tail code was black and orange, and the horizontal bars on the tail were orange, edged with black.*
(Rotramel)

*Right: This close-up of the tail of 154469 illustrates a slight change from the tail markings on the aircraft seen above. The words **GOLDEN WARRIORS** were painted in yellow on the bottom orange band. The colored stars on the rudder were slightly different in both color and style from that seen previously. This Corsair II was used as the squadron's CAG aircraft in late 1974.*
(Centurion Enterprises)

Another nice variation of CAG markings is seen in this 1981 photograph of A-7E, 159968. The special CAG markings were represented by the different colors of the feathers in the headdress.
(Grove)

VA-93 RAVENS

A-7A, 152673, the CAG aircraft for VA-93, was photographed at Misawa AB, Japan, in October 1974. The blue darts on the tail were edged in red, and the flash below the **NF** tail code was red. The horizontal bars on the rudder were (top to bottom) red, yellow, orange, and green. A very attractive shark's mouth was painted on the sides of the intake.
(Taylor via Geer)

Beautiful bi-centennial markings were applied to the CAG aircraft for VA-93, as seen in this landing photograph of 152673.
(Flightleader Collection)

After VA-93 transitioned to the A-7E, 160730 became the CAG aircraft, and it is shown in this photograph that is dated July 3, 1982. The blue and red markings on the tail were slightly different from those seen on the aircraft at the top of this page. **CAG** was stencilled in white on the black fin cap. The horizontal bars on the rudder were also black. (Flightleader Collection)

VA-94 MIGHTY SHRIKES

One of the more colorful Corsair II squadrons during the 1970s was VA-94. A-7E, 159976, was being used as their CAG aircraft at the time this photograph was taken on July 24, 1976. The black **NL** tail code was almost lost in the colorful CAG stripes.
(Flightleader Collection)

Even though the tail code remained **NL,** a different carrier name, **USS KITTY HAWK,** on the tail cone indicates a carrier reassignment. The horizontal bars at the top and bottom of the tail were orange, edged in black. **SHRIKES** was lettered in black on the bottom stripe. The colored stars painted on the rudder were (top to bottom) red, yellow, blue, orange, green, and black.
(Swanberg via Geer)

Another change in markings is illustrated in this 1983 photograph of A-7E, 157516. The letters **CAG** on the fin cap and the stripes and squadron emblem on the tail were orange. (J. Stewart)

VA-97 WARHAWKS

*At left is the right side of A-7A, 153252, as photographed in the early 1970s. The colorful CAG markings provided the background for the **NK** tail code. At right is A-7E, 156872, used as the CAG aircraft in 1975. The light blue tail flash turns to different CAG colors on the rudder. A change in carrier assignment had been made from USS CONSTELLATION to USS ENTERPRISE.*
(Both Flightleader)

A dark blue tail flash was painted on the CAG aircraft seen here in November 1978. The colorful bars on the rudder represented the special CAG markings. (Swanberg via Geer)

*The same aircraft shown above is seen in this photograph that was taken on August 2, 1980, however several changes are illustrated. **USS CORAL SEA** was lettered in black on the tail cone. Notice that the special CAG markings had been repainted horizontally across the rudder from top to bottom.*
(Huston)

VA-105 GUNSLINGERS

A-7E, 158831, was the Corsair II being used by VA-105 for their CAG aircraft when this photograph was taken. The colored chevron on the tail was the special CAG marking, and the photograph below provides a closer look at the colors. The chess figure was black, and was positioned on a white disc that was edged with black. The **AC** tail code on the rudder was green, shadowed in black. *(Curry)*

The same aircraft was photographed in October 1976, and a minor change in markings was made for the bi-centennial celebration. The chess piece was replaced with the red, white, and blue bi-centennial emblem. **THE BEAR** was painted in green on a white rectangle under the cockpit. *(Grove)*

Slightly different CAG stripes were painted on this aircraft as compared to those seen in the photographs above. Only five colored stripes were present, with the bottom brown color being deleted. *(Flightleader)*

*Very colorful CAG markings adorned A-7E, 159970, which is shown here as it taxied out for a flight. A different emblem appeared on the CAG stripes on the tail. The **AC** tail code and the **400** modex on the nose were green, shadowed in yellow. **GUNSLINGERS** on the tail and **VA105** on the tail cone were both lettered in green. The carrier assignment had been changed to the USS JOHN F. KENNEDY, and **CAG SMITH** was in yellow under the cockpit.* (Grove)

*At left is an overall view of the same aircraft awaiting clearance for departure. **USS CARL VINSON** is stenciled in black above **NAVY** on the fuselage. Notice that the in-flight refueling boom is extended. At right is a close-up view of the tail markings. The **NL** tail code, indicating a deployment in the Pacific, is green, shadowed with yellow. The CVW-15 badge is painted over the special CAG stripes.* (Both Grove)

*VA-105 had switched to the tactical scheme by the time this Corsair II was photographed on the deck of the USS FORRESTAL. All markings were a contrasting gray, except for the **400** modex on the nose and the **00** on the fin cap. Both of these were flat black.* (Kinzey)

VA-113 STINGERS

This photograph of A-7E, 158644, was taken in early 1974, and illustrates the markings used on VA-113's CAG aircraft. The entire rudder was painted in various colors for the CAG. The large attractive bee on the tail had been used by VA-113 for many years.
(Spering/A.I.R.)

*Some additional CAG markings had been added to VA-113's paint scheme by the time this photograph was taken in August 1975. **CVW-2** was painted vertically in blue on the rudder. **CAG** was lettered on the blue fin cap in white.* *(Geer Collection)*

*At left is an overall view of A-7E, 158664, which was VA-113's CAG aircraft during the bi-centennial celebration. All of the special CAG markings, **STINGERS**, and the **NE** tail code were painted in red, white, and blue. At right is a close-up view of the markings on the tail.*
(Both Flightleader Collection)

47

The same aircraft, seen at the bottom of the previous page, is shown here as it appeared on February 7, 1978. The unit was still using very colorful markings on their aircraft at this time.
(Rogers)

VA-113 had started using more subdued markings by the time 160730 was photographed at NAS Lemoore, California, on October 21, 1982. The markings were painted in dark gray, to include the smaller bee on the tail.
(Bergagnini)

*A-7E, 160730, was painted overall glossy gray when this photograph was taken in January 1983. A change in tail codes had taken place, and the **NM** was arranged vertically on the rudder.*
(Grove)

VA-146 BLUE DIAMONDS

This photograph shows A-7E, 160866, when it was VA-146's CAG aircraft in September 1981. Blue diamonds were painted on the tail behind the yellow chevron. The **NG** tail code on the rudder and the fin cap were also blue.

(Geer)

By late 1982, CAG markings, in the form of horizontal colored stripes, had been added behind the blue diamonds. These are visible in the photograph at left. **USS KITTY HAWK** was painted above **NAVY** on the fuselage. At right is a close-up view of the markings on the tail. The colored stripes were (top to bottom) red, brown, blue, orange, yellow, black, maroon, and green. Notice the white double nuts on the blue fin cap.

(Both Grove)

All color had disappeared from VA-146's CAG aircraft when it was photographed in August 1983. Markings were in a contrasting gray, except for the **300** modex on the nose, **00** on the fin cap, and the diamonds, all of which were a dark charcoal gray. (Grove)

VA-147 ARGONAUTS

A-7E, 156803, was VA-147's CAG aircraft when it was photographed on December 19, 1970. The large sword on the tail had a black handle and a red outline of the blade. The **NG** tail code was red, shadowed with white, and the double nuts on the fin cap were red. (Spering/A.I.R.)

This Corsair II was photographed as it taxied out for a training mission. A slight change had taken place involving the sword painted on the tail. The red fin cap had **CAG** lettered in white. Notice the small practice bombs attached under the wing. (Grove)

The **NE** tail code on the tail and the carrier name, **USS KITTY HAWK,** are further changes to VA-147's markings. The sword on the tail is slightly smaller than previously seen. **CAPT. DON BAIRD BARON** is stencilled under the cockpit in black. (Grove)

VA-153 BLUE TAILED FLIES

A-7A, 153237, was used at the CAG aircraft by VA-153 in the early 1970s. The vertical tail was light blue, and two white stripes were painted below the **NM** tail code. The colored barbs on the white rudder were the special CAG markings. (Flightleader Collection)

The same aircraft was photographed in 1973, and a change in markings can be seen on the tail. The squadron emblem had been painted between the letters of the tail code, and four small blue stars had been added just aft of the emblem. (MAP)

A different aircraft was used for the CAG when this photograph was taken in 1976. The small stars had disappeared from the tail, and **CAG** was painted in red on the white fin cap. The **NM** tail code was also painted red. (Grove)

VA-155 SILVER FOXES

VA-155's CAG aircraft was A-7B, 154548, when this photograph was taken in April 1973. The stripes and **NM** tail code were dark green, edged with white. (LaBouy via Malerba)

One of the stripes on the tail of 154548 was painted in various colors for the CAG markings. The colored stripe ran from the bottom of the tail to the top of the fin cap. (Bergagnini)

At left is the left side of the same aircraft, which was photographed on May 8, 1977. The squadron badge had been applied to the white rudder. At right is a close-up view of the tail. Notice the carrier name, **USS ROOSEVELT,** that was painted in black above **NAVY.** Shortly after this photograph was taken, the USS FRANKLIN D. ROOSEVELT was taken out of service and scrapped. VA-155, and the rest of her air wing, was disestablished. VA-155 was reestablished as an A-6 Intruder squadron for a short time beginning on September 1, 1987. See page 20 for a photograph of its only CAG aircraft while it was an Intruder squadron. (Both Flightleader Collection)

VA-174 HELLRAZORS

VA-174 is the A-7 RAG (Replacement Air Group) squadron for the Atlantic Fleet. One of its first CAG aircraft was A-7B, 154361, and it is shown here as it appeared in 1968. Yellow diamonds were painted on the black band across the top of the tail. The CAG markings consisted of multicolored bands on the rudder. (Flightleader Collection)

A change in markings for the CAG aircraft is seen in this late 1970 photograph of A-7E, 156876. Notice that the diamonds on the black band across the top of the tail were various colors for the CAG. **COMMANDER LIGHT ATTACK WING ONE** was lettered in black on the fuselage beneath the wing. (Spering/A.I.R.)

Another change in CAG markings is illustrated in this photograph that is dated 1975. The **AD** tail code was painted on a yellow diamond. This diamond was shadowed by multicolored stripes that swept to the front of the tail. **CDR T.C. WATSON JR. COMLATWING ONE** was painted in black under the cockpit. (Sides)

A large wing badge was painted on the tail of A-7E, 157456, which was photographed at NAS Norfolk, Virginia, on May 2, 1976. The **AD** tail code had been moved to a vertical position on the rudder. (Spering/A.I.R.)

53

VA-192 GOLDEN DRAGONS

*Above: A-7E, 156822, was the CAG aircraft for VA-192 when it was photographed at McGuire AFB, New Jersey, on June 13, 1970. Two yellow stripes, edged with black, were on the tail. The rudder was painted in various colors for the CAG markings, and two black nuts were on the yellow fin cap. Also notice the two nuts painted into the **300** modex on the flap.* (Spering/A.I.R.)

Right: This close-up of the same aircraft shows the yellow dragon and CAG's name painted on the front of the aircraft. (Spering/A.I.R.)

A change in markings is seen in this photograph of A-7E, 157530. A large yellow chevron, edged in black, was painted on the tail. The rudder had small chevrons of various colors for the CAG markings. (Flightleader Collection)

*More changes in markings can be seen in this photograph of A-7E, 157546. **USS ENTERPRISE** was lettered above **NAVY** on the fuselage in black, indicating a change in carrier assignment, and the small chevrons had disappeared from the rudder. **CDR JACK READY CVW-11** was stencilled in black on a yellow rectangle below the cockpit.* (Grove)

VA-195 DAM BUSTERS

VA-195 was another Corsair II squadron that had colorful markings during the 1970 time frame, as evidenced by this photograph of 157545. The colored feathers that streamed from the eagle's head to the rudder were the special markings for the CAG. **USS KITTY HAWK** was lettered in black on the tail cone. Two white **00**s were on the fin cap, and each contained a star. The band around the forward fuselage was green, edged in black.
(Flightleader Collection)

A different aircraft was being used for VA-195's CAG aircraft when this photograph was taken on June 28, 1980. Notice the **NH** tail code was completely black, and lacked the white shadowing seen on the aircraft at the top of the page. The carrier assignment had been changed to the USS ENTERPRISE. (Rogers)

The change by VA-195 to the tactical scheme is illustrated on A-7E, 158833. The squadron emblem, which was painted on the tail in light gray, had been changed, and the **NG** tail code was arranged vertically on the fin. **USS RANGER** was painted in black on the tail cone. (Grove)

VA-203 BLUE DOLPHINS

At left is an overall look at A-7A, 153163, which was used as the CAG aircraft by the Atlantic Fleet reserve squadron VA-203. At right is a close-up view which shows the blue panel and fin cap. Small colored dolphins were painted on the two white stripes as the special CAG markings.   (Both Sides)

During the bi-centennial year red, white, and blue markings adorned the tail of the CAG aircraft. Note that this is the same Corsair as shown above. Special colored stripes were painted on the top and bottom of the rudder for the CAG markings.   (Roth)

This photograph of A-7B, 154493, illustrates another CAG aircraft used by VA-203 during the bi-centennial year of 1976. This change in aircraft was necessitated by the squadron's transition from the A-7A to the A-7B. Notice the CAG markings, in the form of colored stripes, painted around the forward fuselage. The bi-centennial colors remained on the tail and extended onto the rudder. *(Meinert)*

The same Corsair II was later painted in the low-visibility scheme with darker gray contrasting markings. The fin cap had a multi-colored pattern for the CAG markings, and **CAG 20** was lettered in dark gray below the fin cap.
(Spering/A.I.R.)

VA-204 RIVER RATTLERS

The second Atlantic Fleet reserve squadron, VA-204, operated this attractive CAG aircraft. An orange stripe, edged with black, and containing **RIVER RATTLERS,** was painted along the bottom of the tail. The black **AF** tail code was shadowed in orange. Four large orange stars were painted on the rudder, and a black **400** modex was stencilled on the orange fin cap.
(Geer)

The markings on this CAG aircraft were experimental markings for VA-204, and were only applied to this aircraft. The black, orange, and yellow rattlesnake was wound through the **AF** tail code. **CVWR-20** was lettered on the tail cone in black.
(Meinert)

This photograph of A-7B, 154547, was taken in April 1982. The special CAG markings included the wing badge on the tail, and the individual squadron badges that surrounded it. **CDR DAVE LAYTON CAG** was stencilled in black below the canopy. (Flightleader)

The right side of 154547 was photographed as it was being checked for a training flight to the bombing range. (Grove)

A change in markings is seen on A-7B, 154488, which was photographed as it taxied out for a flight. A map of the state of Louisiana was painted in orange on the tail, and a black star indicated the location of NAS New Orleans. Not easily seen is the black **AF** tail code in the upper portion of the state. **VA-204 RIVER RATTLERS** was lettered in black on the orange stripe across the bottom of the tail. This is also difficult to see in this black and white photograph. The diamonds on the rudder were various colors for the CAG. (Grove)

VA-205 GREEN FALCONS

The third Atlantic Fleet reserve squadron is VA-205, based at NAS Atlanta, Georgia. A-7B, 154370, was photographed during 1976, and displays both bi-centennial and CAG markings. The barbs on the tail were various colors for the CAG, and a red **76**, surrounded with blue stars, was painted at the bottom of the white rudder. (Flightleader)

At left is an overview of the CAG aircraft as it appeared in early 1978. At right is a close-up of the tail markings, including the green and white Falcon holding a green bomb. The colored barbs on the rudder were spaced differently than those seen in the photograph above. (Both Flightleader)

This view of VA-205's CAG aircraft illustrates the attractive markings that this unit once used. The colored barbs had been painted down the entire height of the rudder. (Grove)

Above: A-7B, 154469, displays various awards on the white rudder. The shape of the bomb being held by the Falcon had been altered from that seen on the previous page. (Wilson)

Left: This close-up illustrates the CAG markings and name that was painted on the nose. The bar below the CAG's name was in various colors. (Wilson)

*Alas, the low-visibility era eventually caught up with VA-205, and they changed to the tactical scheme. Their CAG aircraft, A-7E, 157491, is shown here as it taxied on the flight deck of the USS LEXINGTON in late 1986. A map of the state of Georgia was painted in light gray, with a star and small Falcon indicating the location of Atlanta. The **AF** tail code was painted in the bottom right corner of the state. The markings contrast only slightly with the background color, and are very difficult to see.* (Kinzey)

VA-215 BARN OWLS

A-7B, 154468, was being used by VA-215 at the time this photograph was taken in the early 1970s. The black **AE** tail code indicates an Atlantic deployment aboard the USS FRANKLIN D. ROOSEVELT.
(Flightleader Collection)

By the time this photograph was taken in August 1975, VA-215 was reassigned to the Pacific Fleet and carried **NM** tail codes. The carrier name, **USS ORISKANY,** was lettered in black. The stripes on the rudder were (front to back) red, yellow, blue, orange, and green.
(Geer Collection)

Another change in carrier assignments is indicated by **USS ROOSEVELT** stencilled in black above **NAVY**.
(Peacock)

VA-303 GOLDEN HAWKS

VA-303 is one of the Pacific Fleet reserve squadrons. Their CAG aircraft, A-7A, 153200, was photographed as it was readied for a flight. The rudder was blue, and had colored diamonds for the special CAG markings. Notice the carrier name, **USS RANGER,** on the tail cone. The squadron badge was applied to the tail above the stylized **ND** tail code.
(Flightleader Collection)

A change in squadron markings had taken place by the time this photograph was taken in early 1980. A golden hawk was painted on the light blue tail, and had the black **ND** tail code on it. The colored diamonds on the rudder were the CAG markings.
(Grove)

This photograph of A-7B, 154389, illustrates the markings applied to the left side of the aircraft. (Grove)

VA-304 FIREBIRDS

The second Pacific Fleet reserve squadron is VA-304. One of their first CAG aircraft was A-7A, 153171. The colored shadowing of the stylized **ND** tail code represented the CAG markings. Notice the name, **PAGASUS**, painted in black, and the red flash on the main gear door.
(Flightleader Collection)

After transitioning to the A-7B, 154362 became the CAG aircraft, and it displayed the same CAG markings seen in the photograph above. However, there was no name on the main gear doors. (Flightleader Collection)

A change in markings is apparent in this May 1980 photograph of A-7B, 154362. The **ND** tail code had colored stars along each side as the CAG markings and two orange nuts were painted on the white fin cap. **CAG-30**, which is almost invisible in this photograph, was lettered in yellow and edged in black beneath the tail code.
(Grove)

VA-305 LOBOS

The last Pacific Fleet reserve squadron is VA-305. They had an attractive CAG and bi-centennial aircraft. The **ND** tail code had colored stars painted on both sides as the CAG markings, and the white rudder contained a red and blue design for the bi-centennial celebration. The words, **NAVAL AIR RESERVE POINT MUGU, CALIF.**, were in black on the tail cone, and the **00** on the white fin cap was red and blue. *(Flightleader)*

Left: A close-up view shows the markings on the right side of the tail of 153268. Notice **AIR WING 30** *lettered in black below the tail code.* *(Flightleader)*

More subdued CAG markings appear on this Corsair II from VA-305. Notice the small green Lobo head on the fuselage in front of the wing. **CAPT. JIM HAMILTON CAG** was lettered in white on a green rectangle beneath the canopy. *(Grove)*

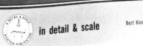

THE FOLLOWING OUTSTANDING VOLUMES IN THE COLORS & MARKINGS SERIES AND THE DETAIL & SCALE SERIES ARE AVAILABLE FROM TAB BOOKS

COLORS & MARKINGS SERIES

Book No. 24525	Vol. 1,	Colors & Markings of the F–106 Delta Dart
Book No. 24526	Vol. 2,	Colors & Markings of the F–14 Tomcat, Part 1 (Atlantic Fleet Squadrons)
Book No. 24527	Vol. 3,	Colors & Markings of the F–4C Phantom II, Part 1 (Post Vietnam Markings)
Book No. 24528	Vol. 4,	Colors & Markings of the F–4D Phantom II, Part I (Post Vietnam Markings)
Book No. 24529	Vol. 5,	Colors & Markings of the A–6 Intruder, (U.S. Navy Bomber & Tanker Versions)
Book No. 24530	Vol. 6,	Colors & Markings of the U.S. Navy Adversary Aircraft (Includes the Aircraft of TOP GUN)
Book No. 24531	Vol. 7,	Colors & Markings of Special-Purpose, C–130 Hercules
Book No. 24532	Vol. 8,	Colors & Markings of the F–14 Tomcat, Part 2 (Pacific Fleet Squadrons)
Book No. 24533	Vol. 9,	Colors & Markings of the A–7E Corsair II (USN Atlantic Fleet Squadrons--Post Vietnam Markings)
Book No. 24534	Vol. 10,	Colors & Markings of U.S. Navy CAG Aircraft, Part I (Fighters, F–8, F–4, F–14)
Book No. 24535	Vol. 11,	Colors & Markings of U.S.A.F. Aggressor Squadrons
Book No. 24536	Vol. 12,	Colors & Markings–MiG Kill Markings From the Vietnam War (U.S.A.F. Aircraft)
Book No. 24537	Vol. 13,	Colors & Markings of the F–4E Phantom II (Post Vietnam Markings)
Book No. 24538	Vol. 14,	F–100 Super Sabre, Part 1 (Regular Air Force Fighter Wing)
Book No. 24539	Vol. 15,	A–7 Corsair II, Part II (Pacific Coast Squadrons)

DETAIL & SCALE SERIES

Book No. 25012	Vol. 2,	B–17 Flying Fortress, Part I (Production Versions)
Book No. 25013	Vol. 3,	F–16 Fighting Falcon (Models A & B)
Book No. 25043	Vol. 4,	F–111 Aardvark (The Aircraft That Bombed Libya) (New Revised Edition)
Book No. 25015	Vol. 5,	F–5E & F Tiger II (USAF & USN Aggressor Aircraft)
Book No. 25016	Vol. 6,	F–18 Hornet (Developmental & Early Production Versions)
Book No. 25017	Vol. 7,	F–4 Phantom II, Part 2 (USAF F–4E & F–4G)
Book No. 25020	Vol. 8,	F–105 Thunderchief (Covers all Fighter--Bomber and Wild Weasel Versions)
Book No. 25018	Vol. 9,	F–14A Tomcat (Su–22 Killer)
Book No. 25019	Vol. 10,	B–29 Superfortress, Part I (Production Versions)
Book No. 25021	Vol. 11,	B–17 Flying Fortress, Part 2 (Derivatives)
Book No. 25022	Vol. 12,	F–4 Phantom II Part 3 (USN & USMC)
Book No. 25027	Vol. 13,	F–106 Delta Dart (Ultimate Interceptor)
Book No. 25028	Vol. 14,	F–15 Eagle (2nd Edition)
Book No. 25025	Vol. 15,	F9F Panther (First Navy Jet to See Combat)
Book No. 25024	Vol. 16,	F9F Cougar (Grumman's First Swept Wing Fighter)
Book No. 25026	Vol. 17,	F11F Tiger (U.S. Navy's First Supersonic Fighter)
Book No. 25023	Vol. 18,	B–47 Stratojet (Production Versions)
Book No. 25030	Vol. 19,	A–10 Warthog (The Tank Killer)
Book No. 25029	Vol. 20,	B–17 Flying Fortress, Part 3 (More Derivatives)
Book No. 25031	Vol. 21,	F–101 Voodoo (2nd Edition)
Book No. 25032	Vol. 22,	A–7 Corsair II (2nd Edition)
Book No. 25033	Vol. 23,	Boeing 707 and AWACS
Book No. 25034	Vol. 24,	A–6 Intruder, Part 1 (Bomber & Tanker Versions)
Book No. 25035	Vol. 25,	B–29 Superfortress, Part 2 (Derivatives)
Book No. 25036	Vol. 26,	F6F Hellcat (The U.S. Navy's Most Important Fighter in World War II)
Book No. 25037	Vol. 27,	B–52 Stratofortress (Covers all Versions)
Book No. 25038	Vol. 28,	AV–8 Harrier, Part I (USMC Versions)
Book No. 25039	Vol. 29,	U.S.S. Lexington (CV–16 to AVT–16)
Book No. 25040	Vol. 30,	F4F Wildcat (Grumman's First "Cat" Fighter)
Book No. 25041	Vol. 31,	F–8 Crusader (Covers All Fighter & Reconnaissance Versions)
Book No. 25042	Vol. 32,	A–4 Skyhawk (US Navy & USMC Versions)
Book No. 25044	Vol. 33,	F–100 Super Sabre (Revised, Expanded Edition)
Book No. 25047	Vol. 34,	USS America (CVA–66, CV–66)
Book No. 25046	Vol. 35,	F–102 Delta Dagger (ADC's First Supersonic Interceptor)

See the Detail & Scale Series and Colors & Markings Series at your local Hobby Shops and Book Stores.

If not available in your area or for more information on the Detail & Scale and Colors & Markings Series books, send $1 for the new TAB Catalog describing over 1300 titles currently in print and receive a coupon worth $1 off on your next purchase from TAB, OR ORDER TOLL-FREE TODAY: 1-800-822-8158.

TAB BOOKS
Blue Ridge Summit, PA 17294-0850